ULTIMATE DINOSAURS
GIANTS FROM GONDWANA

Royal Ontario Museum Press

DAVID C. EVANS · MATTHEW J. VAVREK

w

D1472287

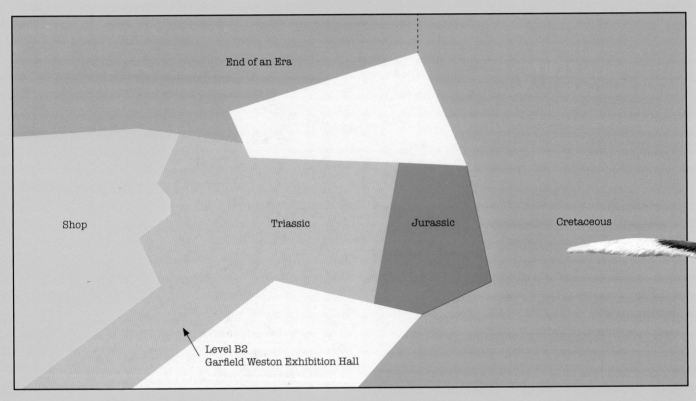

End of an Era

Shop

Triassic

Jurassic

Cretaceous

Level B2
Garfield Weston Exhibition Hall

exhibition floor plan

CONTENTS

5	FOREWORD
6	INTRODUCTION

THE EXHIBITION	14
INTRODUCTORY CONCEPTS	16
PANGAEA AND THE ORIGIN OF DINOSAURS	21
THE JURASSIC DIVIDE	29
CRETACEOUS SPLENDOUR: GONDWANA	37
Africa	39
Madagascar	45
South America	51
LEGENDARY GIANTS: GONDWANA	58
CRETACEOUS LAURASIA	63
THE END OF AN ERA	73
LECTURES AND PROGRAMS	76
SPONSOR/PATRONS	78

FOREWORD

We are very pleased to premiere the ROM-originated blockbuster exhibition *Ultimate Dinosaurs: Giants from Gondwana* . I would like to begin by expressing the Museum's gratitude to Raymond James Ltd., presenting sponsor, as well as to James and Louise Temerty for their generous support of the exhibit. Thanks also to the Louise Hawley Stone Charitable Trust for making this beautiful and authoritative exhibition guide and memento of the visit so affordable to our visitors. The Museum's great donors, patrons, and sponsors enrich the lives of many in their philanthropic initiatives. Special thanks to the Government of Ontario for their inestimable assistance in bringing our treasures of world cultures and natural history to the people of Ontario and beyond.

The job of communicating our great narratives is the business of many of our staff, from the expertise of our exhibition curator, ROM paleontologist Dr. David Evans, and his colleague Dr. Matthew Vavrek, through the design and display proficiency of our exhibition and gallery teams, to our award-winning publishing operations (traditional print and electronic), to the latest and ever-advancing digital technologies to equip us to interface with the e-community of the world. Our interactive storytelling on this exhibition includes innovative iPad stations where dinosaurs are brought to life through Augmented Realty (AR).

Ultimate Dinosaurs: Giants from Gondwana unveils specimens that most of us have never before seen, those of the prehistoric southern hemisphere continent of Gondwana where, on that separate landmass, these dinosaurs evolved differently than their counterparts in the northern hemisphere. The southern colossus *Giganotosaurus* may indeed have been the primeval ruler, even more powerful than the northern *Tyrannosaurus rex*, and the largest land predator that ever lived.

Welcome to the exhibition. Forget all you know about dinosaurs and discover a whole new ancient world. We are certain that the show will be a crowd pleaser for dinosaur aficionados of all ages.

Janet Carding
Director and CEO, Royal Ontario Museum

INTRODUCTION

In North America, and much of the world, images of dinosaurs can be found on everything from breakfast cereal boxes and comic books to T-shirts and television shows. Everyone knows what *Tyrannosaurus rex* looked like and can imagine it hunting the famous three-horned *Triceratops* or the familiar duck-billed dinosaurs. Most of these images that have fuelled our popular appetite for these spectacular creatures depict dinosaurs that are found only in the northern continents, especially North America and Asia. However, the dinosaur fossil finds from the northern hemisphere tell only half of the epic story of dinosaur evolution.

The past two decades have seen a boom in paleontological knowledge of dinosaurs in the southern hemisphere, with many amazing new fossils from South America, Africa, Madagascar, and even Antarctica. These discoveries reveal that the landmasses of the southern hemisphere, sometimes referred to as Gondwana, were home to an almost entirely different set of dinosaurs that evolved on their own unique evolutionary pathways, largely isolated from the north. They include not only new species of dinosaurs but also entirely different families that are rarely found north of the equator.

The most spectacular finds of Gondwanan dinosaurs are derived from the Cretaceous period, towards the end of the Age of Dinosaurs. Huge carnivorous predators such as *Giganotosaurus, Carcharodontosaurus,* and *Spinosaurus* rival *T. rex* for the title of king of the dinosaurs, and titanosaurian sauropods such as

Gondwana: A supercontinent comprising the landmasses of Africa, South America, Australia, and Antarctica, as well as Madagascar and India. It formed when the larger supercontinent of Pangaea broke in two about 150 million years ago, and lasted until about 100 million years ago when it fractured into the individual continents of the southern hemisphere.

Futalognkosaurus and *Argentinosaurus* are considered the largest animals ever to walk the Earth. Only a few decades ago, many of these species were unknown to science and their discovery is changing the way paleontologists view the evolution of dinosaurs.

For much of the Cretaceous, dinosaur faunas from the northern and southern continents were strikingly different. Dinosaur faunas in the northern and southern hemispheres were not, however, always so distinct from one another. At the dawn of the Mesozoic era, during the Triassic period (250 to 200 million years ago), all of the Earth's landmasses were united in a single expansive supercontinent called Pangaea. The first dinosaurs evolved there, and because there were no major geographic barriers preventing animals from moving freely throughout Pangaea, dinosaurs quickly established themselves in every corner of the Earth. This lack of major barriers also meant that different populations (groups of individuals) of dinosaurs could intermingle, and as a result early dinosaur faunas had similar characteristics across the globe.

At the end of the Triassic, a global extinction event occurred—the fifth largest in Earth's history. Although

Pangaea: A giant supercontinent made up of all the continents of the Earth. It formed about 300 million years ago, and eventually broke apart about 150 million years ago to become the two smaller supercontinents of Laurasia (in the north) and Gondwana (in the south).

this extinction was not as devastating as the end-Permian extinction event that paved the way for the appearance of dinosaurs, it nevertheless cleared away many of the most common and sometimes dominant groups of land animals. This extinction opened up ecological niches to new groups, and this time it was dinosaurs that filled much of the empty ecospace. Dinosaurs quickly radiated into the major lineages that would survive for the next 130 million years and grew in body size to become the undisputed dominant land herbivores and predators.

The Jurassic period lasted from 200 to 145 million years ago, immediately following the end-Triassic extinction event. It was during this time that dinosaurs underwent an important diversification, with many new

Ischigualasto Provincial Park, northern Argentina

As this equatorial sea, called the Tethys, expanded, it became increasingly difficult for dinosaurs to move freely between the northern and southern landmasses. Consequently, the sea became a physical barrier to gene flow between the land-living dinosaur populations, and the dinosaurs of Laurasia and Gondwana became increasingly isolated from one another. Over time, this isolation allowed these groups of dinosaurs to evolve on different pathways, and they gained new and unique characteristics related to their northern and southern environments. The separation of Laurasia and Gondwana by the end of the Jurassic was a fundamental divide in the Earth's landmasses that is reflected in a major global division of dinosaur faunas during the subsequent Cretaceous period, and this geographic split persists to this day in many living animals and plants.

The Cretaceous period, from 145 to 65 million years ago and the last in the Mesozoic, hosted the greatest diversity of dinosaurs. It was also during this period of time that many modern groups of animals and plants, such as lizards and angiosperms, began to diversify and push out more ancient groups. It was a time of great transformation on the Earth's surface as Laurasia and

species of dinosaurs appearing as they began to occupy the vacant ecological spaces left in the extinction's wake. The Jurassic period was also a time of great change on Earth. The processes of plate tectonics began to pull the northern and southern portions of the giant supercontinent of Pangaea apart, eventually creating a sea that separated the smaller supercontinent of Laurasia in the north from Gondwana in the south.

Gondwana further divided into what would become today's continents. For the first half of the Cretaceous, the continents of Gondwana remained connected to one another, yet were completely separated from those in the north. Dinosaurs could move relatively unimpeded across all of Gondwana during this time, and the unique character of the Cretaceous Gondwana dinosaur fauna took hold: abelisaurids, spinosaurs, and carcharodontosaurs as the leading predators, with sauropods (notably the titanosaurs) as the dominant herbivores. The Earth's plates continued to move through the Cretaceous, and by the middle of this period, Gondwana itself began to fragment into the individual landmasses we know today. Yet again, the dynamic Earth had begun isolating dinosaur communities through the movement of its crustal plates. In turn, increased isolation promoted the evolution of unique animals (including dinosaurs) and plants on these newly independent landmasses.

The astounding diversity of Cretaceous Gondwanan dinosaurs is best represented by the fossil record of three landmasses: Africa, Madagascar, and South America. The Cretaceous dinosaurs of Africa have come from countries primarily in the northern part of the continent such as Egypt, Morocco, Niger, and Tunisia. In particular, Niger has produced some of the most spectacular fossils of dinosaurs such as *Suchomimus*, *Nigersaurus*, and *Ouranosaurus*. Although many of the areas where African dinosaurs are found today are now within the Sahara Desert, during the Cretaceous much of this area would have been covered by waterways and mangrove swamps.

Madagascar has also produced an incredible array of Cretaceous dinosaurs—including some of the best preserved skeletons ever found. Dinosaurs such as the sauropod *Rapetosaurus* and the bulldog-faced carnivore

Gene Flow: The movement of genetic information between two or more populations (groups) within a species. If gene flow becomes restricted between populations, there is a higher chance that the populations will evolve into different species.

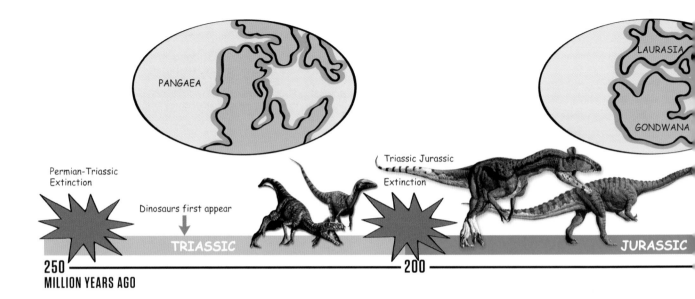

PANGAEA

LAURASIA

GONDWANA

Permian-Triassic
Extinction

Dinosaurs first appear

Triassic Jurassic
Extinction

TRIASSIC

JURASSIC

250 ——————————————————— 200 ————————————

MILLION YEARS AGO

Majungasaurus called the island home. There are also bones from a wide variety of other animals that lived alongside dinosaurs in their ecosystem—a bewildering diversity of crocodilians, including the plant-eating *Simosuchus*, small mammals, birds, and even a giant "devil" frog called *Beelzebufo* that was large enough to eat baby dinosaurs. All of these animals lived at the very end of the Age of Dinosaurs, at the same time that the great tyrannosaurs and horned dinosaurs flourished in North America.

10

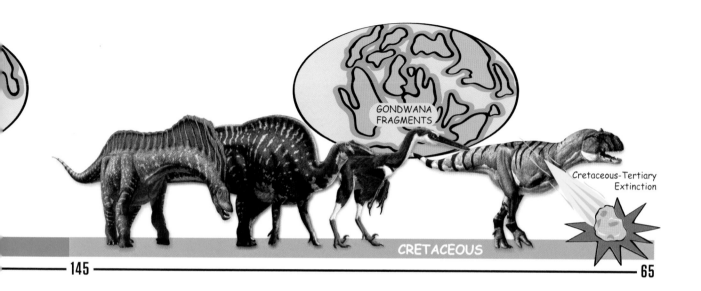

GONDWANA FRAGMENTS

Cretaceous-Tertiary Extinction

CRETACEOUS

145 — 65

In South America, the dinosaur record spans almost the entire Cretaceous period. More dinosaur fossils have been recovered in South America than on any other continent in the southern hemisphere. As a result, a great diversity of species is known from South America, most of which come from Patagonia in Argentina. The rocks of Patagonia have produced the fossils of the meat-eating *Carnotaurus*, with paired horns over its eyes and tiny arms, and the giant *Argentinosaurus*, the largest animal known to have ever walked the Earth.

Paleontologists excavate fossils of *Futalognkosaurus* in Patagonia, Argentina.

By the close of the Cretaceous, the southern continents had been largely separated from the northern continents for about 80 million years—more time than separates humans from the last non-avian dinosaurs and the entire Age of Mammals. This fundamental separation between north and south helped promote the evolution of very different ecosystems in these two places, with each region having a unique flora and fauna that evolved in isolation. By the Late Cretaceous, the southern landscape was dominated by the long-necked sauropods feeding high and low, while in the north the horned ceratopsian dinosaurs and large duckbills were common herbivores. Likewise, the carnivores in these two areas were also very different. In the south, carcharodontosaurs such as the giant *Giganotosaurus* gave way to abelisaurids such as *Carnotaurus* and *Majungasaurus*, while in North America tyrannosaurs such as *Tyrannosaurus rex* were on top of the food chain. Although the giant carnivores *T. rex* and *Giganotosaurus* were both bipedal predators, they are very different beasts, products of lineages that diverged tens of millions of years apart. *Giganotosaurus* had a relatively elongate skull and more blade-like teeth, making it better suited to slicing meat off prey. On the other hand, *T. rex* had a much more robust skull and thick, spike-like teeth that could be used for crushing bones.

Although at one time the mass extinction at the end of the Cretaceous was thought to have wiped out all the dinosaurs, we now know that they found their way to the modern world as birds, which are the direct descendants of small, feathered theropod dinosaurs. And since the end of the Cretaceous, plate tectonics has continued to shape the distribution of the continents on the Earth. Australia broke away from Antarctica, and India slammed so violently into Asia it raised the Himalaya Mountains to their record heights. The shape and position of the continents continue to influence patterns of biodiversity to the present day, with mammals now the major players. The isolation of Australia and Madagascar played an important role in their unusual animal and plant life. Today, volcanic eruptions and earthquakes remind us that the Earth's continents continue to move. The dramatic interplay between the dynamic Earth and the evolution of life will continue millions of years into the future.

> **Extinction:** The eradication of all of the individuals of a species. In the past this was the result of natural causes (e.g., those that precipitated the demise of the dinosaurs). Currently, humans are the principal cause of the extinction of species.

THE EXHIBITION

INTRODUCTORY CONCEPTS

THE EVOLVING EARTH:
THEORY OF PLATE TECTONICS

Plate tectonics is the scientific explanation of why the continents move. Based on the types of minerals present, the Earth can be divided up into a series of layers, starting from the solid inner core, through the outer core and mantle, with a rocky outer crust at the surface. The surface of the Earth that we can see, called the lithosphere ("sphere of rock"), is made up of distinct individual plates, and each plate can move independently of the others. The various continents are supported by these plates. A process called mantle convection drives much of the movement of the plates. Below the solid outer layer of the Earth is a zone of the mantle called the aesthenosphere ("sphere of weakness"). The aesthenosphere is not solid rock but is somewhere between a solid and a liquid—think plasticine or toffee. Heat within the Earth drives convective currents within the aesthenosphere, which are created when hotter parts rise and cooler parts sink, much like the convective currents that create winds. The currents in the aesthenosphere, however, move along very slowly, at only a few centimetres per year. These currents move across the entire lower surface of the plates, and this slow yet very powerful force is enough to move the plates and their continents above.

crust

mantle

core

lithosphere aesthenosphere

PIECING IT ALL TOGETHER: ALFRED WEGENER AND PANGAEA

The theory of plate tectonics is a relatively new idea in science, and up until the mid-twentieth century many thought that the continents of the Earth were static. Alfred Wegener (1880–1930), a German scientist, first had the idea that the continents may not have always been arranged as they are now, and that they were once connected in one single landmass. He devised the Theory of Continental Drift in the early part of the twentieth century after noticing numerous instances of identical fossil species on continents now separated by vast oceans. For example, fossils of the land-dwelling synapsid *Lystrosaurus* had been found in both South America and Africa, on opposite sides of the Atlantic Ocean.

During Wegener's time, the prevailing idea was that animals moved between continents over vast, submerged land bridges. Wegener noticed that the Atlantic coastlines

of Africa and South America appeared to fit almost perfectly together. He theorized that instead of animals and plants moving between continents, it was the continents themselves that moved after the fossils had been formed. Wegener presented his ideas in 1915 with the publication of his book *The Origin of Continents and Oceans*. The idea was promptly dismissed by most geologists and paleontologists. But in the decades that followed his death, new information on the Earth's crust led to general acceptance of plate tectonics in the 1960s. Plate tectonics relates how the structure of the Earth leads to slowly moving plates and vindicates Wegener's idea of drifting continents.

EVOLVING LIFE:
CHARLES DARWIN AND BIOTIC EVOLUTION

Evolution is fundamental to explaining the history and diversity of all life, which descended from a single common ancestor more than 4 billion years ago. This definition encompasses population-level microevolution, such as changes in gene frequency in a population from one generation to the next, to macroevolution, which encompasses the formation of new species and how they are related to each other. Charles Darwin (1809–1882) introduced the concept of evolution in his 1859 publication *On the Origin of Species*. Biological evolution is heritable descent with modification, and it is the process of change in life over generations. Evolution within populations is often driven by the processes of natural selection, whereby individuals with beneficial, heritable traits have more offspring and pass along their characteristics, while those individuals that lack these

traits have fewer offspring, eventually removing these traits from the population. That adaptively advantageous changes are heritable is central to the mechanism of natural selection.

When a population of a species becomes isolated via the formation of geographical barriers that limit or eliminate gene flow (interbreeding), locally different selective forces can pull each population in different evolutionary directions. This process is termed vicariance, the process by which the geographic range of a population is split apart by the formation of a physical barrier. For the land-living dinosaurs, the fragmentation of Pangaea formed seas that acted as these barriers.

NATURAL SELECTION

One of the major mechanisms of evolution is a process called natural selection. Natural selection is the process whereby individuals within a population that are better adapted to their environment tend to survive and produce more offspring. Although this basic principle of natural selection is relatively simple, it is often misunderstood.

For example, imagine a population of moths. In this population, there is variation in certain traits, such as the colour of the moths. Because colour has a genetic basis, blue moths have blue babies and brown moths have brown babies. Not all individuals of a population will survive to reproduce, but individuals with certain beneficial traits may survive more often than others. For example, birds find brown moths to be less desirable than blue moths. As a result, they eat blue moths more often than they eat brown moths. In this case, brown colouration is an advantageous trait, with more brown moths surviving to have babies. Over time, brown moths will become more and more common, until eventually all the moths in the population may be brown.

TRIASSIC

PANGAEA AND THE ORIGIN OF DINOSAURS

About 250 million years ago, the largest mass extinction ever, called the Permian-Triassic Extinction, annihilated more than 90 per cent of all species on Earth. Entire families of organisms disappeared, and whole ecosystems were devastated. What was devastating for some lineages, such as the trilobites, however, was an opportunity for others. The extinction of so many species meant that ecological roles that had once been filled were suddenly empty and available to be filled by the survivors. Among the groups that diversified in the aftermath of this extinction were the ancestors of the dinosaurs

The first evidence of dinosauromorphs (dinosaurs and their close relatives) consists not of fossilized bones but rather fossilized footprints. In 2010, fossilized trackways from a dinosauromorph dating from just after

260 MYA

The skeleton of *Prestosuchus*, a rauisuchian crurotarsan, an early relative of crocodiles

the Permian-Triassic Extinction were found in Poland. These early trackways were made at the same time that ecosystems were recovering from the effects of the mass extinction, and it is possible that without the ecological opportunities that arose, the earliest dinosaurs may never have evolved.

The first fossilized bones of true dinosaurs come from Ischigualasto Provincial Park in Argentina. Rocks found here date from the earliest part of the Late Triassic, about 230 million years ago. Dinosaurs fossilized within these rocks include *Herrerasaurus*, *Pisanosaurus*, and *Eoraptor*.

A COMPETITIVE WORLD

Dinosaurs are not the only animals found in the rocks of Ischigualasto. In fact, dinosaur fossils are relatively uncommon compared to many of the other groups of animals located there. Animals such as the herbivorous rhynchosaurs, an array of cynodonts (mammal relatives),

Cynognathus, a close relative of mammals, was a common Triassic animal found across much of the southern hemisphere.

and the giant carnivorous rauisuchian *Saurosuchus* (a distant relative of crocodiles) are found together with these earliest dinosaurs, and often in a greater abundance and variety.

In the face of this competitive environment, early dinosaurs inhabited more limited ecological roles. They were generally small and bipedal, often predators or omnivores, and they had to compete with other similarly sized and even much larger animals for resources. Because the landmasses were still joined into a single supercontinent, Pangaea, the early dinosaurs spread quickly across the globe. For the same reason, they were of a similar character with limited differentiation or diversity.

250 MYA

EORAPTOR

Eoraptor was a small, bipedal dinosaur that lived during the Triassic, about 228 million years ago. *Eoraptor* had two different types of teeth, with those in the back more serrated and curved, while those in the front were more leaf-shaped, suggesting that it ate a variety of foods and was an omnivore. Its exact position on the family tree of dinosaurs has been controversial. It was once thought that *Eoraptor* was a theropod, but new research indicates that it is in fact the oldest sauropodomorph dinosaur, a member of the lineage that leads to the giant long-necked sauropods.

ETYMOLOGY: "dawn plunderer"

LENGTH: 1 m

WEIGHT: 15 kg

FORMATION: Ischigualasto

AGE: Late Triassic

HERRERASAURUS

Of the earliest dinosaurs that we know from Ischigualasto, *Herrerasaurus* was one of the biggest. Like its contemporary *Eoraptor*, it was bipedal and had a relatively long tail. However, it was a carnivorous theropod that may have preyed on some of the other dinosaurs that lived in the same area, such as *Pisanosaurus* or *Eoraptor*. *Herrerasaurus* had a robust skull and a jaw that had a joint in the middle—the intramandibular joint—which made the jaw more flexible to help protect the teeth from breaking when they were embedded in struggling prey.

ETYMOLOGY: "Herrera's lizard"

LENGTH: 4 m

WEIGHT: 200 kg

FORMATION: Ischigualasto

AGE: Late Triassic

240 MYA

THE TRIASSIC-JURASSIC MASS EXTINCTION

Although the first dinosaurs appeared after a mass extinction at the beginning of the Triassic, their rise to dominance was assisted by another mass extinction 50 million years later at the end of the Triassic. The Triassic-Jurassic extinction profoundly affected life on land and in the oceans. Although smaller than the Permian-Triassic and end-Cretaceous mass extinctions, it was still a devastating event that affected life globally. During a relatively short period of time, a large number of animal groups became extinct, including most of the common types of land vertebrates. The specific causes of this extinction remain a mystery. Several explanations for this event have been proposed: climate change, sea-level fluctuation, massive volcanic eruptions, and even an asteroid impact similar to the one associated with the end-Cretaceous mass extinction that killed off all the non-avian dinosaurs. Whatever happened, numerous life forms vanished. But, perhaps by luck alone, the dinosaurs survived.

Several groups of common Triassic animals were virtually wiped at the end of the Triassic

The asteroid that created the Manicouagan crater is one suggested cause for the Triassic-Jurassic extinction

230 MYA

JURASSIC
THE JURASSIC DIVIDE

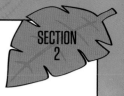

The mass extinction event left the surviving dinosaurs with many new ecological opportunities. In the absence of many of their former competitors, the different groups of dinosaurs proliferated. Not only did the dinosaurs survive the Triassic-Jurassic mass extinction, but they also radiated into many new forms and took over the planet. Dinosaurs now thrived, growing much larger, and many new groups evolved.

At the beginning of the Jurassic, the continents remained largely joined, providing a continuous land route along which species could move. These connections between the continents also meant that the types of dinosaurs found across the globe were generally similar to one another. However, during the Jurassic a seaway opened up along the centre of Pangaea, near the equator. By the end of the Jurassic, this seaway divided Pangaea into two separate landmasses in the north and south, called Laurasia and Gondwana, respectively.

220 MYA

CRYOLOPHOSAURUS

ETYMOLOGY: "frozen crested lizard"

LENGTH: 6.5 m

WEIGHT: 450 kg

FORMATION: Hanson

AGE: Early Jurassic

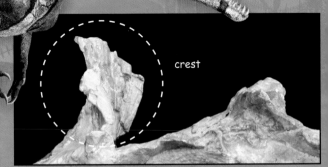

crest

Cryolophosaurus was the first named dinosaur from Antarctica. Its fossils were excavated by paleontologist William Hammer from Augustana College in Illinois, with the help of Philip Currie of the University of Alberta and others. Excavating *Cryolophosaurus* from Antarctica was no easy task. The site where it was found was only 650 kilometres from the South Pole, in the Transantarctic Mountains, which protrude through the ice sheets of Antarctica to expose Triassic and Jurassic rocks. Even though the research team worked in the middle of the Antarctic summer, temperatures rarely rose above -20°C. The fossils were also in very hard rocks, and the only way to get them out was through the use of rock saws, jackhammers, and, in some cases, dynamite.

A large portion of the dinosaur has been recovered, including parts of the skull that show that it had a prominent "pompadour" crest across the top of its head—lending the animal its name. Although the

Excavating fossil bones of *Cryolophosaurus* in Antarctica

specimen that was found is estimated to be 6.5 metres long, features of its skeleton suggest that it wasn't fully grown, so as an adult it would have been even larger. After three visits to the site, researchers have recovered additional fossils of different species of dinosaurs and other animals that provide evidence of a rich Antarctic ecosystem about 190 million years ago.

210 MYA

MASSOSPONDYLUS

The long-necked, small-headed dinosaur *Massospondylus* belongs to a group called the sauropodomorphs, which were the dominant terrestrial plant eaters throughout the Jurassic period. Numerous *Massospondylus* fossils have been found in South Africa, including multiple complete skeletons of different growth stages. Even embryos in eggs have been found—the oldest dinosaur embryos ever recovered. These reveal that *Massospondylus* started life on all four legs, and became bipedal as it grew to adulthood. Recent fieldwork by Robert Reisz of the University of Toronto and David Evans of the Royal Ontario Museum has resulted in the discovery

ETYMOLOGY: "massive vertebrae"

LENGTH: 5 m

WEIGHT: 160 kg

FORMATION: Upper Elliot

AGE: Early Jurassic

Golden Gate Highlands National Park, South Africa

of multiple complete nests of eggs that constitute a dinosaur nesting site. The astonishing series of 190-million-year-old nests has given scientists the first detailed look at dinosaur reproduction early in their evolutionary history. The site documents the antiquity of nesting strategies such as group nesting and site fidelity, two behaviours that were previously known only from much later in the dinosaur record.

200 MYA

A FUNDAMENTAL DIVIDE: PANGAEA SPLITS IN TWO

During the Jurassic, great forces from within the Earth slowly divided the uppercontinent Pangaea into two landmasses, Gondwana in the south, and Laurasia in the north. By the beginning of the Cretaceous period (145 million years ago), total separation of Laurasia and Gondwana had occurred, with the two landmasses separated by a newly formed sea called Tethys, which was centred around the equator. This separation set the stage for dinosaurs to follow different evolutionary paths, and the ecosystems of the south and north became quite distinct from one another. Further break-up of Gondwana continued through the rest of the Cretaceous and into the Age of Mammals. This fragmentation of Gondwana resulted in further isolation of dinosaur populations in the Late Cretaceous, from which the greatest diversity of dinosaurs is known.

A skeleton of the well-known Jurassic dinosaur *Stegosaurus*

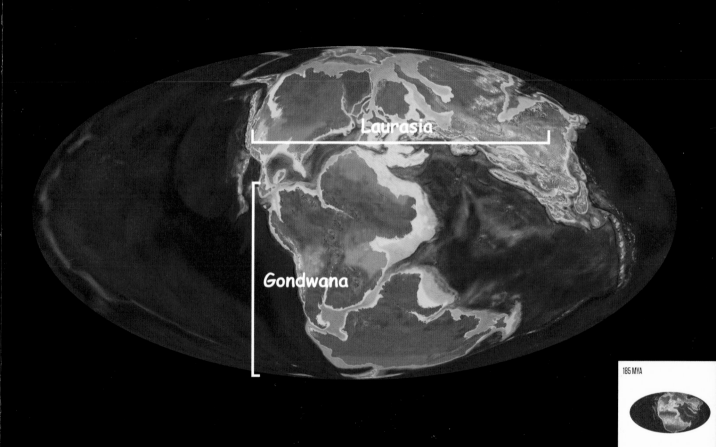

Laurasia

Gondwana

185 MYA

CRETACEOUS
CRETACEOUS SPLENDOUR:
THE GREAT GONDWANA DINOSAURS

Cretaceous Period / 145–65 million years ago

Throughout the early part of the Cretaceous, the continents of the southern hemisphere were connected to one another, allowing different dinosaur groups to evolve and disperse across the southern supercontinent of Gondwana. Because Gondwana was completely isolated from the northern supercontinent of Laurasia, it prevented the mixing of dinosaurs between the northern and southern hemispheres. Eventually, the individual landmasses of Gondwana began to slowly break up into many of the individual southern-hemisphere continents that we know today. By the end of the Age of Dinosaurs, most of the different landmasses were separated from one another, further isolating the different dinosaur groups from one another, and leading to unique faunas on each of the continents.

170 MYA

AFRICA

The Cretaceous dinosaurs of Africa are best known from finds concentrated in the northern Sahara Desert region. Several spectacular dinosaur finds come from an area in Niger called Gadoufaoua, including fossil skeletons of *Ouranosaurus* and *Suchomimus*. Gadoufaoua is a term in the local Touareg language that means "the place where camels fear to go," and today it is an incredibly hot and dry area. During the Cretaceous, however, northern Africa looked very different. Large parts of the desert were covered by lush environments such as mangrove swamps, and others were flooded by inland seas. The Gadoufaoua beds preserve an amazing diversity of dinosaurs and other animals such as huge lungfish and numerous crocodiles, including the giant 12-metre-long *Sarcosuchus*, which may have preyed upon dinosaurs drinking at the water's edge.

Although the most complete Cretaceous dinosaur skeletons discovered from Africa so far come from Gadoufaoua, fossils have also been found in other African countries. For example, remains of the giant theropod

Above: Skull and neck of *Malawisaurus*, a titanosaur sauropod
Opposite: The skull of *Carcharodontosaurus*, a giant theropod from Africa

Carcharodontosaurus have been found in the Kem Kem beds of Morocco, while the sauropod *Malawisaurus* was named after the country of Malawi where it was found. Egypt, Sudan, and Libya have also produced dinosaur remains.

160 MYA

NIGERSAURUS

teeth

ETYMOLOGY: "Niger lizard"

LENGTH: 9 m

WEIGHT: 4000 kg

FORMATION: Elrhaz

AGE: Early Cretaceous

Nigersaurus was a rebbachisaurid sauropod with a relatively short neck and a long tail. This species was described by paleontologist Paul Sereno and colleagues in 1999. Some of its bones, however, were first found by Philippe Taquet in the 1970s. The most unusual feature of this dinosaur is its bizarre head. All of its teeth were concentrated in a row at the front of the jaws. These teeth, which were continually worn down and replaced, were packed conveyor-belt style. It had about 50 teeth in a row on both the top and bottom jaws, and each tooth had about nine more replacement teeth behind it, meaning that it had hundreds of teeth in its mouth at any given time. The amazingly well preserved braincase and inner ear cast, together with the shape of the skull and neck, tell scientists that *Nigersaurus* held its head and mouth facing downwards, suggesting that it was a low browser that fed on plants at ground level.

SUCHOMIMUS

In 1997, a team led by Paul Sereno first came across the fossilized remains of *Suchomimus* in the Sahara desert of Niger. The team managed to recover almost two-thirds of the skeleton, making it one of the best-known spinosaurs. Like other spinosaurs, *Suchomimus* has a skull with a long snout similar in shape to that of a crocodile, hence its name "crocodile mimic." Characteristics of its skull and teeth indicate that this animal, along with its larger sail-backed cousin *Spinosaurus*, were probably fish eaters that fed on the gigantic fish species thriving in the rivers and lagoons at the time. The skulls of spinosaurs and crocodiles are examples of convergent evolution, the acquisition of a similar biological trait in unrelated lineages in response to similar ecological challenges, in this case, catching fish.

ETYMOLOGY: "crocodile mimic"

LENGTH: 11 m

WEIGHT: 3000 kg

FORMATION: Elrhaz

AGE: Early Cretaceous

150 MYA

OURANOSAURUS

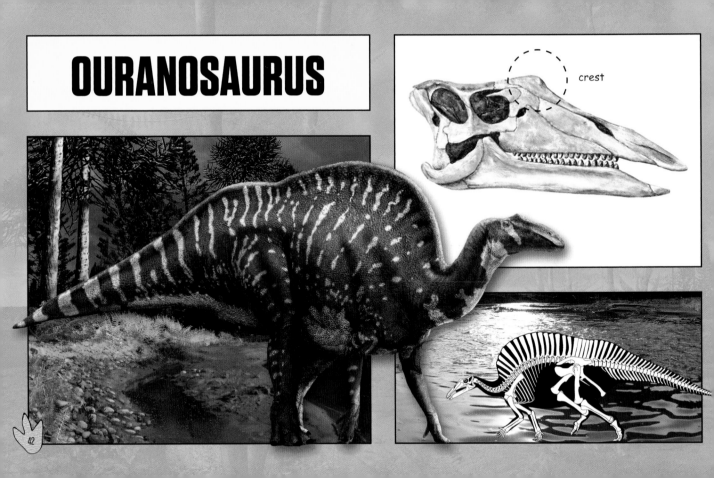

crest

Ouranosaurus was a large ornithopod dinosaur closely related to the well-known *Iguanodon* from Europe. It was a medium-sized plant-eating animal, attaining a length of about 7 metres and weighing around 3,000 kg in life, about as much as a rhinoceros. The most distinctive feature of *Ouranosaurus* is the row of elongated vertebral spines along its back, which form a "sail" that stands more than 1.5 metres high. The form and function of this sail have been debated by paleontologists. Some have suggested that it may have acted as a large radiator to regulate body temperature, but the most widely accepted idea is that it functioned as a flamboyant "billboard" to signal to other members of its species. Another idea is that the sail was in fact a hump that may have stored fat and other nutrients, like a camel's hump. *Ouranosaurus* had an unusual skull, the nostrils positioned far back

ETYMOLOGY: "brave (monitor) lizard"

LENGTH: 7 m

WEIGHT: 3000 kg

FORMATION: Elrhaz

AGE: Early Cretaceous

Paleontologists uncover the sail of *Ouranosaurus* in the sand of the Sahara Desert

on the snout and a small crest situated above the eyes. It also had a thumb spike on the hand, like that of *Iguanodon*, although the spike was relatively smaller.

140 MYA

MADAGASCAR

Our knowledge of the dinosaurs of Madagascar comes primarily from one area in the northwestern end of the island, where the Late Cretaceous Maevarano Formation is found. The rocks of this formation were deposited by large rivers, about 70 million years ago, after the island had separated from the rest of Gondwana. During this time, the climate of the area was similar to the seasonal and semi-arid climate of Madagascar today.

Historically, fossils excavated in Madagascar consisted of poorly preserved material. Since 1993, a large-scale research program undertaken by paleontologist David Krause and his co-workers has uncovered spectacular new finds of almost complete fossil skeletons which record the astounding diversity of animals on the island shortly before the end of the Age of Dinosaurs. The Maevarano Formation includes the fossil remains of birds, mammals, crocodilians, dinosaurs, and even a giant frog.

120 MYA

MAJUNGASAURUS

arm
and hand

Majungasaurus was an abelisaurid theropod that lived on the island of Madagascar at the end of the Age of Dinosaurs, approximately 70 million years ago. *Majungasaurus* is known from many excellently preserved fossils from many individuals, and is the best-known member of the abelisaur lineage. Its skeleton has been completely documented, and the anatomy of its head has been reconstructed in detail using CT-scan technology. *Majungasaurus* had a thick, ornamented skull with a strange dome-like projection over its forehead. Like its relative, *Carnotaurus*, it had a relatively short, deep face with powerful jaws that could deliver a strong, slashing bite. Strangely, its arms and hands were incredibly short, and they could not have been used for hunting. Tooth-marked bones provide direct evidence of what *Majungasaurus* ate.

ETYMOLOGY: "Mahajanga lizard"

LENGTH: 8 m

WEIGHT: 2000 kg

FORMATION: Maevarano

AGE: Late Cretaceous

Fossil bones of the large sauropod *Rapetosaurus* are commonly found with tooth marks that match the teeth of *Majungasaurus*. Bones of *Majungasaurus* with *Majungasaurus* tooth marks, however, also show that it was a cannibal, at least some of the time. It is not clear if it was an opportunistic feeder, eating kin that were already dead, or if it killed other members of its species for food.

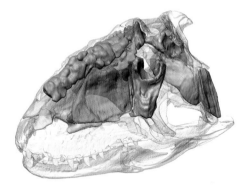

A 3-D model of a *Majungasaurus* skull showing internal soft tissue regions

115 MYA

MASIAKASAURUS

dental structure

A medium-sized theropod, *Masiakasaurus knopfleri* is probably best known for being named after the singer Mark Knopfler from Dire Straits, "whose music inspired expedition crews." The most distinctive feature of this dinosaur is its dental structure. Instead of having teeth pointing upward and downward, the teeth at the front of its mouth project directly forward. The function of this strange dentition is uncertain. It may have allowed *Masiakasaurus* to catch small prey such as fish or other small animals.

ETYMOLOGY: "vicious lizard"

LENGTH: 2 m

WEIGHT: 20 kg

FORMATION: Maevarano

AGE: Late Cretaceous

One of the most amazing discoveries made in Madagascar to date is not a dinosaur, but a bizarre crocodilian called *Simosuchus*. *Simosuchus* is unlike any other crocodile relative living today because it was a herbivore. It had leaf-shaped teeth adapted for eating plants. Similar teeth are found in modern herbivorous lizards, as well as in many plant-eating dinosaurs, such as ankylosaurs. *Simosuchus* gets its name from its pug-like face, very different from the long-snouted faces of crocodiles today. *Simosuchus* was small, about the size of a large dog, and it had armour in its skin to help protect it from predators.

ETYMOLOGY: "pug-nosed crocodile"

LENGTH: 1.5 m

WEIGHT: 35 kg

FORMATION: Maevarano

AGE: Late Cretaceous

SIMOSUCHUS

dental structure

105 MYA

SOUTH AMERICA

South America has produced the vast majority of dinosaur fossils from Gondwana. Although dinosaur fossils have been known from South America for more than a century, it is through the recent work of Argentinian paleontologists that this impressive diversity has come to light. The Cretaceous rocks of Patagonia are particularly rich and have provided some of the most remarkable fossils ever found. Dinosaurs of South America come from a larger variety of rock formations than those of either Africa or Madagascar. This variety means that many of the best known South American dinosaurs would have lived in different environments at different times. For example, the sail-necked sauropod *Amargasaurus* lived in the Early Cretaceous, whereas the horned meat-eater *Carnotaurus* is from the final stage of the Late Cretaceous. Although their skeletons were both found in Patagonia, they would have lived at least 50 million years apart.

100 MYA

CARNOTAURUS

horns

skin impression

Carnotaurus is known for a large pair of horns on the top of its head, which lead its discoverers to give it a Latin name that translates as "meat-eating bull." The horns were probably used to communicate visually with members of its own species, either to attract mates or to compete with rivals. The almost complete skeleton of *Carnotaurus* was first described by famous paleontologist José Bonaparte. The skeleton was found with preserved skin impressions across most of its right side, from the neck to the tail. The extent of these impressions, which consist of scales of varying sizes arrayed in different patterns, allows scientists to suggest what this dinosaur would have looked like. Because of this fossil record, we know that some theropods related to *Carnotaurus* did not have feathers, unlike more advanced theropods and their cousins, birds.

ETYMOLOGY:	"meat-eating bull"
LENGTH:	7 m
WEIGHT:	1500 kg
FORMATION:	La Colonia
AGE:	Late Cretaceous

Argentine paleontologist Fernando Novas with the skull of *Carnotaurus*

The skeleton of *Carnotaurus* provided the first complete look at an abelisaurid. Abelisaurids are a group of short-faced, tiny-armed theropods found almost exclusively in Gondwana. *Carnotaurus* is related to *Majungasaurus* from Madagascar, as well as other abelisaurids from India and Africa. It lived during the latest part of the Cretaceous (Campanian-Maastrichtian), just before the end of the Age of Dinosaurs.

90 MYA

AUSTRORAPTOR

The long-snouted *Austroraptor* lived about 70 million years ago in what is now Patagonia. At more than 5 metres in total length, it is one of the largest known members of the dromaeosaur family, the group of carnivorous theropods that includes the agile *Velociraptor* of *Jurassic Park* fame. It was relatively closely related to the lineage of dinosaurs that led to modern day birds, and probably had feathers of some kind. However, *Austroraptor* had very small arms in relation to its large size, and it would not have been able to fly. Although it was a dromaeosaur, its arms were so short that they were probably not long enough to catch prey. *Austroraptor* is a unenlagiine, a member of a group of dromaeosaurs with long faces and conical teeth known only from South America and Africa.

ETYMOLOGY: "southern thief"

LENGTH: 5 m

WEIGHT: 170 kg

FORMATION: Allen

AGE: Late Cretaceous

BUITRERAPTOR

Buitreraptor was a small, turkey-sized dinosaur that lived about 95 million years ago in what is now Argentina. Although the original fossil did not show any preserved feathers, several of its close relatives have been found with feathers, meaning that *Buitreraptor* also likely possessed feathers of some kind. The skeleton of *Buitreraptor* is the most complete of any dromaeosaur from Gondwana. Its discovery and scientific analysis led to some interesting questions about how and when birds began to fly. Although *Buitreraptor* was flightless, it appears that some of its close relatives may have been able to fly. This implies that flight may have evolved twice among dinosaurs, once in birds and once among the Gondwanan dromaeosaurs.

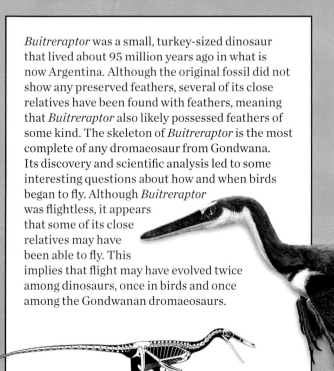

skull

ETYMOLOGY: "vulture raider"

LENGTH: 1.5 m

WEIGHT: 3 kg

FORMATION: Candeleros

AGE: Late Cretaceous

85 MYA

AMARGASAURUS

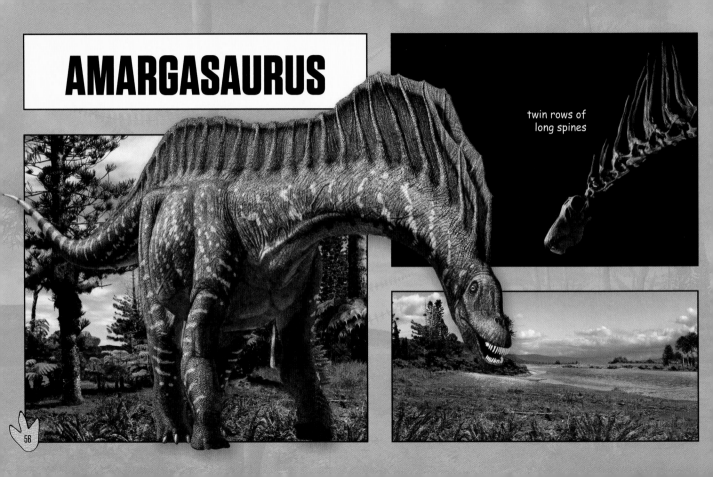

twin rows of
long spines

Amargasaurus is an unusual sauropod found in Patagonia (Neuquen Province), Argentina. Its skeleton was found in 1984 by a team led by José Bonaparte. Although somewhat small compared to many other sauropods, it was more than 10 metres in length. This dinosaur is known for the twin rows of long spines (up to 50 cm) running down its neck. The spines may have been connected to skin, giving the creature a distinctive frill along the length of its body. Like the flamboyant head crests and horns of other dinosaurs, this garish outgrowth may have been used to attract mates or to compete with rivals, and may have even been brightly coloured to make it more visible. *Amargasaurus* lived during the Early Cretaceous (Barremian-Aptian), or about 120 to 130 million years ago. A herbivore, it used its peg-like teeth to strip low-lying vegetation.

José Bonaparte with a mounted skeleton of *Amargasaurus*

ETYMOLOGY: "Amarga lizard"

LENGTH: 10 m

WEIGHT: 4000 kg

FORMATION: La Amarga

AGE: Early Cretaceous

75 MYA

LEGENDARY GIANTS:
THE TITANOSAURS OF GONDWANA

The titanosaurs include the largest land animals ever to have walked the Earth. Titanosaurs are a group of giant long-necked, plant-eating dinosaurs that were common in the Cretaceous of Gondwana. They are characterised by a stocky build and wide stance compared to other sauropods. Some of the largest known species, including *Futalognkosaurus* and *Argentinosaurus*, measured more than 30 metres long, greater in length than two school buses, and may have weighed as much as 70,000 kg when alive—more than the weight of ten African elephants. Although adult sauropods were enormous, baby sauropods hatched out of eggs no bigger than a soccer ball. The microstructure of many dinosaur bones reveals that most, including the giant titanosaurs, grew at rates equal to or exceeding that of mammals. The large sauropods have been estimated to have gained up to 3 kilograms per day during their fastest growth phase!

Argentinosaurus, the largest land animal ever found

Futalognkosaurus was one of the largest animals ever to have walked this planet, and it would have dominated the landscape. Not only was it one of the largest animals, its skeleton is also the most complete of any of the large long-necked dinosaurs that we know. Because so much of the skeleton was found, we are able to learn so much more about how this animal both lived and died. The skeleton of *Futalognkosaurus* was found in the high plains of Patagonia, Argentina, and was named by a team led by Argentinian paleontologist Jorge Calvo in 2007.

A front view of the hips of *Futalognkosaurus*

FUTALOGNKOSAURUS

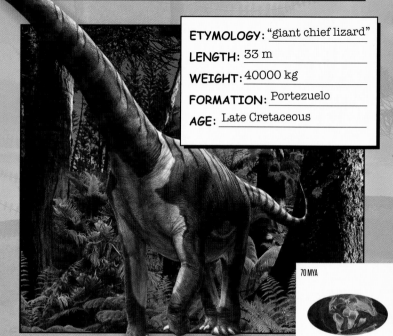

ETYMOLOGY: "giant chief lizard"

LENGTH: 33 m

WEIGHT: 40000 kg

FORMATION: Portezuelo

AGE: Late Cretaceous

70 MYA

GIGANOTOSAURUS

Giganotosaurus may have been the largest land predator ever to have lived. It is the largest carnivorous dinosaur from Gondwana known from a complete skeleton—the best specimen is missing only its arms and feet. The skeleton suggests that *Giganotosaurus* was approximately 13 metres long and stood about 3 metres tall at the hips. Although similar in size to *Tyrannosaurus rex*, its skull was longer and less robust. It had blade-like teeth well suited to slicing meat, rather than the more bone-crushing teeth and jaws of *T. rex*. *Giganotosaurus* was the apex predator in South America about 95 million years ago, where it hunted titanosaurs and small ornithopod dinosaurs, and shared its world with the small carnivore *Buitreratpor*. *Giganotosaurus* is similar in size to *T. rex* in Laurasia

ETYMOLOGY: "giant southern lizard"

LENGTH: 13 m

WEIGHT: 6000 kg

FORMATION: Candeleros

AGE: Late Cretaceous

Rodolfo Coria with mounted skeleton of *Giganotosaurus*

and *Carcharodontosaurus* and *Spinosaurus* in Africa. Their closeness in size suggests they may be near the maximum possible size for any land predator, and nothing since has rivalled this size.

65 MYA

CRETACEOUS LAURASIA:
A NORTHERN PERSPECTIVE

The continents of the northern hemisphere, including North America, Asia, and Europe remained connected for much of the Cretaceous period. By the Late Cretaceous, the dinosaurs of Laurasia and Gondwana had been isolated from each other for more than 50 million years. Over those 50 million years, evolution took very different paths in the two hemispheres, leading to very different and distinctive worlds in the north and south.

In the Late Cretaceous, when sea levels were high, large portions of the northern continents were flooded by inland seaways, further separating and isolating the lands from one another. North America was divided in two by a large sea that stretched from the Arctic Ocean to the Gulf of Mexico, while Europe consisted of a series of islands. These seas also helped to preserve a staggering diversity of Laurasian dinosaurs within the great fossil fields of western North America, including the badlands of Alberta, and in central

60 MYA

Asia. Although many low-lying areas were flooded, a land bridge in Beringia connected Asia and North America. It is because of this bridge that dinosaurs found in these two areas are often closely related.

Ornithischians, including large-bodied duck-billed dinosaurs such as *Edmontosaurus* and the horned dinosaurs such as *Triceratops* and its kin, were the most abundant plant-eaters in Laurasia, and the great tyrannosaurs and dromaeosaurs were the predators that lived off them. This contrasts with the giant plant-eating sauropods and meat-eating abelisaurs that lived in Gondwana at the same time.

The fossil beds of western North America have been intensely worked by paleontologists for more than 100 years, and the abundant dinosaur fossils have found their way into museums across the world. As a result, paleontologists know a great deal about Cretaceous ecosystems. This allows detailed comparisons of the Cretaceous dinosaur faunas from the northern and southern hemispheres during this time.

Above: Badlands in Dinosaur Provincial Park, Alberta, Canada
Opposite: *Lambeosaurus*, a duck-billed dinosaur from the Late Cretaceous of North America

THE DUCK-BILLED DINOSAURS

Large-bodied ornithopod dinosaurs were the dominant plant-eaters in many Laurasian ecosystems. The duck-billed dinosaurs, or hadrosaurs, are the best-known representatives of this group. These animals evolved broad, toothless bills for cropping plants, and a complex dental battery made up of hundreds of teeth that they used to chew their food. In fact, their efficient way of processing plants was likely an important factor in their success. Duck-billed dinosaurs evolved into a diverse array of species distinguished by the forms of their skulls. There were two groups of duck-billed dinosaurs, the generally flat-headed hadrosaurines, and the flamboyantly crested lambeosaurines, which are characterized by hollow crests that contained their nasal passages. The hadrosaurs first evolved in Laurasia, where they were ubiquitous, but some hadrosaurs made it to South America via a brief connection between North America during the latest Cretaceous.

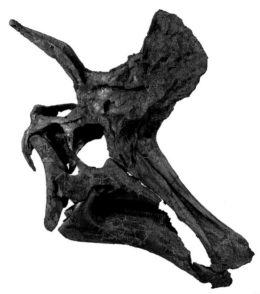

50 MYA

PARASAUROLOPHUS

Parasaurolophus lived in western North America from about 77 to 72 million years ago. Its fossils have been found from Alberta to New Mexico. *Parasaurolophus* is the most recognizable duck-billed dinosaur because of its long, tube-shaped crest that projects from the back of its skull. This crest, more than a metre long, was actually hollow. It surrounded greatly elongated tubes of the nasal passage. It is thought that this crest served both as a visual display and as a resonating chamber to modify sounds that were made to communicate with other members of its own species.

ETYMOLOGY: "near crested lizard"

LENGTH: 8 m

WEIGHT: 3000 kg

FORMATION: Dinosaur Park

AGE: Late Cretaceous

EDMONTOSAURUS

Edmontosaurus lived in North America in the last 5 million years of the Cretaceous, and its remains have been found over much of the western half of the continent. More fossils have been found of *Edmontosaurus* than of any other duck-billed dinosaur; more than 25 skulls have been unearthed to date and many complete skeletons are known. *Edmontosaurus* is understood from the numerous and massive bone beds containing the remains of thousands of individual animals, which suggest that this dinosaur lived in herds at least part of the year—like wildebeest today. Fossils found in Alaska above the Arctic Circle indicate that *Edmontosaurus* would have lived there year round and was able to tolerate cold conditions for at least part of the year.

ETYMOLOGY: "Edmonton lizard"

LENGTH: 13 m

WEIGHT: 4000 kg

FORMATION: Horseshoe Canyon

AGE: Late Cretaceous

45 MYA

THE HORNED DINOSAURS

The horned dinosaurs range from small-bodied forms such as *Psittacosaurus* to the massive *Triceratops,* which weighed more than an elephant. The first horned dinosaurs are known from the Late Jurassic of Asia, but their remains are common in both Asia and North America throughout the Cretaceous. Many of the smaller, early members of the group such as *Protoceratops* are known in abundance from Asia. The characteristic horns and frills of *Triceratops* and its ceratopsid relatives may have been used for visual displays or even combat between members of a herd, perhaps to identify mates or battle for dominance. Large-headed ceratopsids, such as *Chasmosaurus*, *Centrosaurus*, and *Triceratops*, occur predominantly in the Late Cretaceous of North America, but a member of this group, *Sinoceratops*, was recently discovered in China. No ceratopsian fossils have been found in Gondwana, suggesting that horned dinosaurs evolved uniquely in the northern hemisphere.

Centrosaurus, a horned dinosaur from the Late Cretaceous of Canada

TRICERATOPS

Triceratops, the most famous horned dinosaur, is one of the best known in terms of fossil material. Even though more than a hundred skulls of this iconic dinosaur have been collected and are in museums around the world, research is revealing new information about how *Triceratops* grew. The horns of juveniles curled backwards and their short frills had tall triangular projections adorning them. As the animal grew, the horns straightened out and re-curved forwards, the frill elongated, and the triangular projections were reduced to low, rounded nubs. This research suggests that *Torosaurus*, previously thought to be a huge, distinct species that lived alongside *Triceratops*, actually represents more mature individuals of *Triceratops*.

ETYMOLOGY: "three-horned face"

LENGTH: 9 m

WEIGHT: 7000 kg

FORMATION: Hell Creek & Lance

AGE: Late Cretaceous

35 MYA

THEROPODS

Theropods took on a diverse array of forms, from large and small meat-eaters, to toothless omnivores, and strange plant-eaters. The great diversity of theropods is well represented in the Cretaceous of Laurasia. The ostrich dinosaurs, the ornithomimids, and the strange oviraptorosaurs had toothless jaws paired with long, clawed arms. Therizinosaurs were bizarre creatures that walked almost upright and had leaf-shaped teeth, which suggest a herbivorous diet. That all of these groups are found only in Laurasia provides further evidence of the isolation of these huge regions caused by movements of the Earth's plates. The earliest tyrannosaurs first appeared in the Late Jurassic of Asia and quickly spread through Laurasia, where they became the apex predators of the Late Cretaceous.

A skull of the small theropod dinosaur *Deinonychus*

TYRANNOSAURUS

Tyrannosaurus rex is known from numerous good skeletons found across the western interior of Canada and the United States. For decades, *T. rex* was the undisputed king of the dinosaurs, as it was thought to be the largest meat-eater ever to evolve on land. Its supremacy has been challenged in the last 20 years with the discovery of *Giganotosaurus* in South America and new finds of *Carcharodontosaurus* and *Spinosaurus* in Africa. Although which of these giants was the biggest remains unsettled, without question *T. rex* had the strongest jaws; they could bite down with more force than that of any other animal before or since. The teeth got more robust with growth, and by the time *T. rex* reached full size, they were thick spikes well suited for splintering the bones of its prey, which may have even included other *T. rex*—new evidence suggests that, like *Majungasuarus*, *T. rex* was a cannibal.

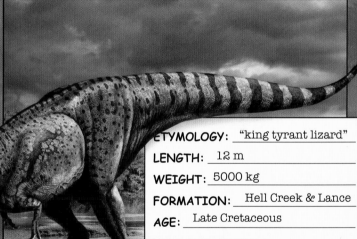

ETYMOLOGY: "king tyrant lizard"

LENGTH: 12 m

WEIGHT: 5000 kg

FORMATION: Hell Creek & Lance

AGE: Late Cretaceous

30 MYA

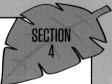

THE END OF AN ERA

The dinosaurs dominated the world for more than 140 million years. From their small beginnings on Pangaea in the Triassic, they radiated and evolved into a great diversity of body shapes and sizes, from the largest of sauropods weighing more than 50,000 kg to small feathery birds. By the end of the Cretaceous, most of the southern continents had the outlines familiar to us today (although in slightly different locations). The climate was very warm, sea levels were higher than at any point in the planet's history, and vast areas of the continents were covered by warm, clear, shallow seas. If we could look more closely at the landscape, we would recognize many of the plants and animals, including flowering plants, pine trees, small mammals, lizards, snakes, and birds.

The dinosaurs of the north and south, passengers on these drifting lands, had been isolated from one another for millions of years, allowing evolution to shape them in unique ways.

10 MYA

In the north, the dinosaur fauna was dominated by tyrannosaurids, such as *Tyrannosaurus rex*, as well as by the herbivorous ceratopsians and hadrosaurs. In the south, the giant predatory carcharodontosaurids such as *Giganotosaurus* gave way to abelisaurs as the last top predators, living in a landscape filled with long-necked sauropods.

The great extinction 65 million years ago at the end of the Cretaceous, however, wiped out most of this diverse and dynamic group. The only dinosaurs to survive this cataclysm were birds. The shape of the continents continued to influence biodiversity to the present day, now with mammals as the major players. Mammals inherited the large-bodied ecological niches vacated by the dinosaurs.

Above: Living birds are the direct descendents of dinosaurs
Opposite: A large meteorite impact may have caused the K-T mass extinction.

PRESENT

LECTURES
DINOSAUR HUNTERS

This fall the whole family will get the ultimate dinosaur experience with the opportunity to meet some of the world's foremost dinosaur hunters, or what we like to call paleontologists. Interested in dinosaurs? Want to grow up to be a paleontologist? These are definitely the people you want to talk to!

   Each afternoon will feature a lively meet-and-greet with a well-known dinosaur expert followed by a family friendly talk* in our theatre.

*Ticket purchase required for talk.

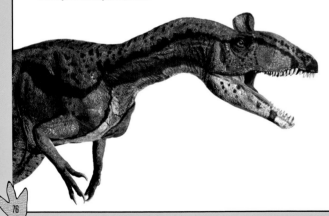

Free Meet-and-Greet (1–2 pm)
followed by a ticketed lecture (2–3 pm):

Sunday, September 16:
In Search of African Dinosaurs:
New Discoveries from South Africa and Sudan
Dave Evans, Associate Curator of Vertebrate Paleontology, ROM

Sunday, October 7:
Dinosaurs of Argentina: New Research and Discoveries
Phil Currie, Professor and Canada Research Chair in Dinosaur Palaeobiology, University of Alberta

Sunday, October 21:
Hold the date—we are working on confirming a very special Dinosaur expert—worth the wait!

Date to be Confirmed:
Baby Dinosaurs of Gondwana
Catherine Forster, Associate Professor of Biology, George Washington University

Saturday, December 8:
The Bizarre and Marvelous Dinosaurs of Madagascar
David Krause, Distinguished Service Professor, Anatomical Science, Stony Brook University School of Medicine

Paleo Playground

In addition to all of the awesome Dinosaur Hunters, each Dino Day will be loaded with free special events and activities for families, kids and dinosaur lovers of all ages.

Dig for dinosaur bones, make your own fossil, build a dinosaur, watch as paleontolgists prepare real fossils for display. You will also be able to see and even touch real specimens from our dinosaur vault. Crafts, costumes, special film screenings, and scavenger hunts will fill your day with dino delight.

All events and activities (except lectures) are included with Museum admission.

Please note that programming is subject to change.

NEW! FRIENDS OF PALAEONTOLOGY

This brand new Friends group offers an exciting opportunity to become involved in a community of individuals and families who are fascinated by fossils (including dinosaurs') and the history of life, and who are interested in supporting palaeontology and palaeontological programming at the ROM. To find out more or to join visit **rom.on.ca/members/fp.php**

Ticket prices for lectures:

Adults $12
Students (15–25 years with valid ID) $10
Children (4–14 years) $8

ROM Members $10
Students (15–25 years with valid ID) $8
Children (4–14 years) $6

•• ROM MEMBERSHIP ••

Join today and apply your admission toward the cost of your new membership. Plus, you'll receive one month free!

Offer available on new membership purchases only, on day of visit. Admission receipt must be presented at time of purchase. Does not apply to Royal Patrons' or Young Patrons' Circle membership levels.

RAYMOND JAMES LTD.

PRESENTING SPONSOR OF
ULTIMATE DINOSAURS:
GIANTS FROM GONDWANA

A key value for the people of Raymond James is giving back into the communities in which they live and serve clients across Canada. One area of focus is support for cultural and arts activities that enrich, educate and inspire. This is why Raymond James chose to partner with the ROM as presenting sponsor of *Ultimate Dinosaurs: Giants from Gondwana.*

This unique exhibition combines cutting edge technologies to enable visitors to look back into time and watch these fascinating creatures come to life. It is a one-of-a-kind educational exhibition that will entrance and enlighten visitors at every stage.

Meticulously curated, designed, and produced by the ROM, *Ultimate Dinosaurs: Giants from Gondwana* is a superb reflection of the combined strengths of the Museum's in-house research and curatorial teams, led by paleontologist Dr. David Evans.

"We congratulate all of the tremendously talented employees and volunteers at the ROM for creating this important exhibition," says Paul Allison, Chairman and CEO Raymond James Ltd. "We look forward to experiencing this spectacular journey and are proud to partner with the ROM to create fascinating and fun memories for dinosaur explorers of all ages."

RAYMOND JAMES®

JAMES AND LOUISE TEMERTY

EXHIBIT PATRON
**ULTIMATE DINOSAURS:
GIANTS FROM GONDWANA**

Avid museum-goers, James and Louise Temerty discovered the ROM when they moved to Toronto from Montreal over 35 years ago.

Over the past three decades, they have been immersed in the life of the Museum, actively involved with the ROM Board of Governors and several committees including the Renaissance ROM Campaign Executive, the Donor Relations and Recognition Task Force, the Nominations Committee, the Finance Committee and the Museum Advancement Committee.

James and Louise remain deeply connected with the ROM today and continue to support the Museum in many ways. They are members of the Currelly Society and dedicated annual supporters of the Royal Patrons' Circle (RPC). The ROM proudly named the James and Louise Temerty Galleries of the Age of Dinosaurs in recognition of their tremendous volunteer leadership and generosity during the Renaissance ROM Campaign.

Most recently, they have generously supported the ROM's blockbuster show *Ultimate Dinosaurs: Giants from Gondwana*, the biggest and best dinosaur exhibition ever mounted in Canada, as Exhibit Patrons.

Published by the Royal Ontario Museum with the generous support of the Louise Hawley Stone Charitable Trust. The Stone Trust generates significant annual funding for the Museum, providing a steady stream of support that is used to purchase new acquisitions and to produce publications related to the ROM's collections. The Louise Hawley Stone Charitable Trust was established in 1998 when the ROM received a charitable trust of nearly $50 million—the largest cash bequest ever received by the Museum—by its long-time friend and supporter, the late Louise Hawley Stone (1904–1997).

Royal Ontario Museum 100 Queen's Park Toronto, Ontario M5S 2C6
www.rom.on.ca

Library and Archives Canada Cataloguing in Publication
Evans, David, 1980–

   Ultimate dinosaurs : giants from Gondwana / David Evans, Matthew Vavrek ; illustrations by Julius Csotonyi.
Issued also in French under title: Prodigieux dinosaures.
Guide to an exhibition held at the Royal Ontario Museum.

ISBN 978-0-88854-491-9

   1. Dinosaurs—Exhibitions. I. Vavrek, Matthew J., 1982–
II. Royal Ontario Museum III. Title.

QE861.4.E93 2012   567.9   C2012-902935-1

David C. Evans is an Associate Curator in Vertebrate Paleontology and oversees dinosaur research at the Royal Ontario Museum. He is also a cross-appointed Assistant Professor in the Department of Ecology and Evolutionary Biology at the University of Toronto.

Matthew J. Vavrek is an Assistant Curator at the Royal Ontario Museum, and his research involves the evolution and ecology of terrestrial ecosystems through time.

Project Manager: Glen Ellis; Editors: Andrea Gallagher Ellis, Sheeza Sarfraz; Design: Tara Winterhalt; Production: Claire Milne

Cover Photo: *Giganotosaurus*, Julius Csotonyi

Photos: Lucille Betti-Nash, pp. 47, 48, 49; Ron Blakey, p. 35; Brian Boyle, pp. 22, 34, 36–37, 65, 68, 70, 39; Sarah Burch, p. 46; Louie Psihoyos, Corbis, p. 57; Julius Csotonyi, pp. 3, 14–15, 20–21, 23–25, 28–33, 36–37, 40–42, 44–46, 48–52, 54–56, 58–59, 62–63, 66–67, 69, 71–73, 76–77; Phil Currie, pp. 31; 52; David Evans, pp. 32, 55; Dorling Kindersley, Getty, p. 42; Win Initiative, Getty, p. 4; David Hardy, p. 75; Scott Hartman, p. 24–25, 31–32, 40–42, 46, 48–49, 52, 54–55, 57, 61, 66–67, 71; Hélène Vallée, iStock, p. 74; Rob MacMahon, pp. 52–53, 59–61; Peter Makovicky, p. 30; Raul Martin, p. 26; Claire Louise Milne, pp. 3, 11, 12, 16, 19, 21, 29, 37, 73; Ian Morrison, p. 53; National Aeronautics and Space Administration (NASA), p. 27; Dianne Scott, p. 32; Matthew Vavrek, p. 40; Larry Witmer, p. 47

Printed and bound in Canada by Transcontinental Interglobe, Beauceville East, Quebec

The Royal Ontario Museum is an agency of the Government of Ontario.

MIX
Paper from responsible sources
FSC® C011825